GRIFFITH

OBSERVATORY

GRIFFITH OBSERVATORY

A Celebration of its Architectural Splendor

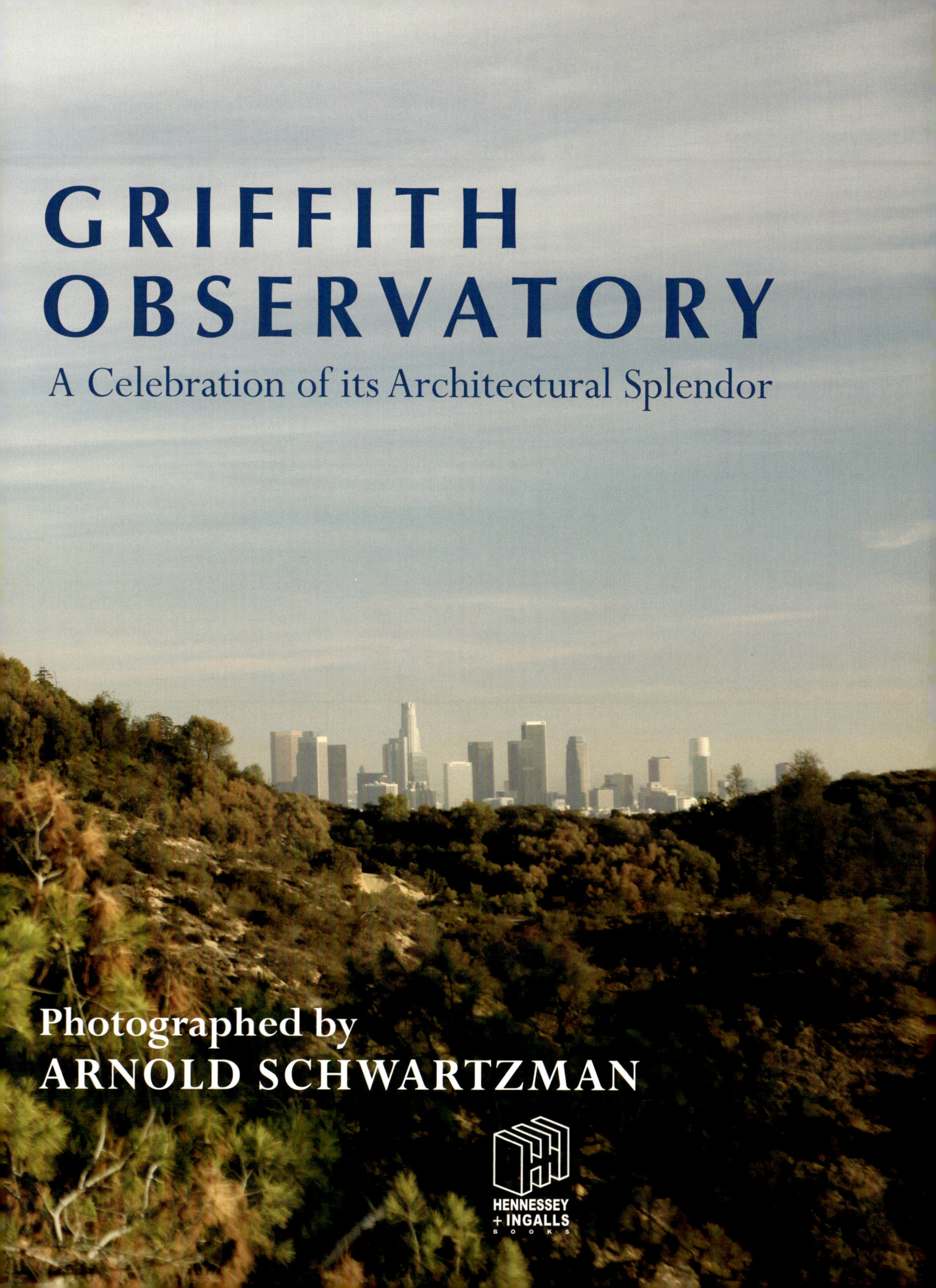

Photographed by
ARNOLD SCHWARTZMAN

HENNESSEY + INGALLS BOOKS

ENTRANCE

Contents

Preface 9

Foreword 10

Introduction 13

Griffith J. Griffith 16

John C. Austin 19

Astronomers Monument 41

The Monument Sculptors 45

The Muralists
Hugo Ballin / Anthony Heinsbergen 67

Acknowledgments 142

Opposite: Detail of Observatory's bronze north entrance door.

Preface

Griffith Observatory's Expanding Universe. What *is* that brilliant gleam on Mount Hollywood? Installed in 2006, the beautiful new copper dome of the Griffith Observatory signals the renovation and expansion of the world-famous Los Angeles landmark.

With its vast history in our great City, the Observatory has created a name for itself. It closed in 2002 for renovations that included a major expansion of exhibit space. When the Observatory reopened to the public on November 3, 2006, it retained its original Greek Revival/Art Deco exterior.

The $93 million renovation, paid in part by a public bond issue, restored the building, as well as replaced the aging planetarium dome. The building was expanded underground with completely new exhibits, a café, gift shop, and the new Leonard Nimoy Event Horizon theater.

Today, the Observatory reminds Angelenos and visitors alike of this complex project. After the restorations, in a presentation to the Los Angeles City Council, Dr. E.C. Krupp, Griffith Observatory Director, explained, "We are thinking cosmically, acting locally, and visible worldwide".

When the Observatory reopened in early 2006, it became the spotlight of Los Angeles, similar to the Getty Center and Walt Disney Concert Hall.

Griffith Observatory is one of the city's leading educational facilities, and hosts more than 25,000 school children each year on field trips. It is often their first introduction to astronomy and the wonders of the night sky. I'm not sure what is more thrilling – marveling at the universe revealed by the Observatory's huge refractor telescope or taking in the view of the Los Angeles basin and the Pacific Ocean from the Observatory's roof deck. There's nothing like it from any other vantage point in the city.

The Observatory is truly one of the great treasures of Los Angeles, and it has helped generations of lifelong learners make sense of the cosmos since 1935.

Tom LaBonge
Councilmember, Fourth District
City of Los Angeles

Griffith Observatory, with downtown Los Angeles framed by the slopes of Mount Hollywood, viewed from Western Canyon Road.

Foreword

There is no place on the planet like Griffith Observatory – maybe no place in the universe.

Residents of Los Angeles and visitors from around the world have been making the astronomical pilgrimage up Mount Hollywood to the Observatory since 1935 for the uncommon perspective Griffith Observatory confers through its vistas, its programs, and its instruments, particularly the Zeiss 12-inch twin refractor, through which more people have looked than any other telescope on earth. In a remarkable convergence of style, purpose, and location, the Observatory is everyone's stepping stone to the stars in southern California.

Visible from throughout the Los Angeles basin, Griffith Observatory is a fundamental element of the landscape. It is an emblem of southern California, and casual eavesdropping daily confirms that visitors instinctively sense its purpose, its dignity, and its accessibility.

Because the Observatory is so familiar, however – with enough major film credits to qualify it for a star on Hollywood Boulevard – its singularity in the realms of public monuments and public astronomy is not what comes to mind. People instead become engaged with the details of their personal experiences at the Observatory.

For example, in the W.M. Keck Foundation Central Rotunda, graced with Hugo Ballin's eye-catching, idiosyncratic, unfashionable, and unparalleled mythological and scientific murals, visitors are mesmerized by the Foucault Pendulum, which demonstrates the primary motion of the earth, its rotation, in real time. You can encounter Foucault Pendulums in museums, planetaria, and other public buildings worldwide, but few people experience more than one and they have no idea how distinctive Griffith Observatory's pendulum is. It is, in fact, the classiest, most memorable Foucault Pendulum on earth.

The live and theatric programming in the Observatory's Samuel Oschin Planetarium develops narrative and theme on behalf of the meaning of astronomy to people. The Observatory's

planetarium shows are like no other.

The Big Picture in the Richard and Lois Gunther Depths of Space is the biggest astronomical image in the world, a real data set that puts people – like every instrument in the Observatory – eyeball to the universe.

The daily and seasonal shifts of the light on the domes, ramparts, terraces, stairways, side-walks, and lawns inevitably capture the heart. Visitors routinely collect colorful sunsets, compelling eclipses, planet conjunctions, passing comets, and other celestial events at this strikingly romantic junction of earth and sky. Nowhere opens the curtains on the celestial stage quite like Griffith Observatory.

It is hard to match the rich range of visual, astronomical, and emotional responses prompted in people by Griffith Observatory and difficult to document them, but Arnold Schwartzman has let his eye turn every corner of the building, follow the Observatory's every line and curve, and absorb every image on behalf of this fetching book. He has caught the Observatory in one stunning architectural moment after another, and in doing so, he elegantly reminds us all how artfully Griffith Observatory connects us to the cosmos.

Dr. E.C. Krupp
Director, Griffith Observatory

South prospect of Griffith Observatory, viewed from Barnsdall Park, Hollywood Boulevard.

Sidewalk terrazzo medallion depicting Griffith Observatory. Clifton's Brookdale Cafeteria, 1936. 648 S. Broadway, Los Angeles.

Introduction

From this noble height one may behold the glories of the fairest land God gave to his children, and no man standing so near to the stars...

— GRIFFITH J. GRIFFITH

Reaching for the Stars. Just months shy of my birth, Griffith Observatory opened its doors to the public. While the Klieg lights of Hollywood were illuminating the galaxy of stars on the boulevard below, dark clouds were gathering over my London home. Batteries of searchlights pierced the night sky, eclipsed by the silhouettes of enemy aircraft.

On September 7,1940 our home suffered a direct hit by the Luftwaffe's bombs, resulting in my evacuation to the countryside where I attended the village school. The teacher, aware of my inability to catch up with the much older pupils, presented me with postage stamps and illustrated cigarette cards to keep me occupied, which lead to my pursuit of the visual arts.

Left: British cigarette card "Do you know what a comet is?". Imperial Tobacco Company, 1933.

On graduating from art school in 1955, I awaited my call-up for my two-year military service. A longtime film buff (at age fifteen I had been an assistant projectionist at our local cinema), I whiled away my time by going to the movies. It was a slow year for the British film industry; meanwhile it was a stellar year for Hollywood. While eagerly awaiting the release of the latest James Dean film, I became engrossed in a couple of Hitchcock thrillers, a Danny Kaye jest, some enchanting musicals, plus a romance, *Love is a Many-Splendored Thing*, set high on a windy hill in Hong Kong.

Right: A salute to Hollywood's film industry. Sidewalk terrazzo medallion depicting a film camera operator. Clifton's Brookdale Cafeteria, 1936.

Little did I know that a few months later I would be shipped out to that very same hill overlooking the British colony, then transported to the foot of another peak, Mount Kamak-San, close to South Korea's DMZ. One cold and dismal evening, I made my way to a large Quonset hut to watch that night's film: *Rebel Without A Cause*. The film's audio was drowned out by the thundering noise made by my fellow combatants stamping their boots on the hut's floor boards in an effort to keep warm. Closing my wire-framed parka hood to a thin letterbox slot, I finally got to view this classic James Dean film.

Its climactic scenes were filmed on location at a magnificent Art Deco building, which I later discovered was Griffith Observatory, situated on Mount Hollywood – some six thousand miles away from Mount Kamak-San. This was my first sighting of what was to become some sixty years on the subject of this book.

In 1978, I followed in the footsteps of other Britons who moved to Southern California, such as architect John D. Parkinson, as well as industrialist Colonel Griffith J. Griffith and John C. Austin, respectively the donor and the designer of the Observatory.

My dream of walking along Hollywood's star-studded boulevard finally came to fruition in the form of an invitation from the legendary graphic designer and filmmaker Saul Bass, to become his studio's design director. On my arrival, I soon discovered that John Milton's lines, "The riches of Heaven's pavement, trodden gold," referred not to those golden stars that paved Hollywood Boulevard but to another firmament that could be viewed from an acropolis high up in the Hollywood hills: the Griffith Observatory.

Following the Observatory's opening in 1935, a like-minded cinephile wrote to the *Los Angeles Daily News:*

I don't want to squint my eye
And look at planets in the sky
Or nebulae, I'd just as soon
Not see the craters in the moon;
Or Venus, Jupiter or Mars,
There is just one, of all the stars
The sight of which would bring
  me joy –
I wanna see the Myrna Loy.

One of my first assignments at the Bass studio was to design a poster for the Girl Scouts of America. My concept was to photographically replicate the new Girl Scouts logo. I recruited a group of local scouts and took them to Griffith Park where I found a suitable location for my shoot. From this spot I was able to glimpse Griffth Observatory in the distance.

Some time later I took the opportunity to make a pilgrimage up to the Observatory. As I approached the man-made plateau where the symmetrical triple-domed building stands, I was impressed by the magnificent surrounding vista of mountains on one side, the Hollywood hills and its famed sign on another, and the sprawl of Los Angeles and the ocean beyond.

Before I even reached the building, I was awe-struck by the large and handsome Astronomers Monument that stands before it. Forty feet high and made of cast concrete, it is surmounted by a cast-bronze armillary sphere. The monument was a project of the Depression-era federal Public Works of Art Project. Standing in each recess of the star-shaped configuration are the statues of six significant astronomers, each created by prominent local sculptors, including George Stanley, best known for his model of the Oscar statuette (one of which sits proudly on my bookshelf).

As I examined and grew to know the exterior of the Observatory, I noticed that the architect had adapted delicate touches of Greek Revival design to his Art Deco palette, such elements as fluting from the Doric order, acanthus leaves, and the Greek zigzag design known as the meander pattern. I later discovered that these additions were originally proposed by astronomical consultant Russell W. Porter as a connection to the park's Greek Theatre.

I was impressed by the bold bronze astronomical graphic symbols, which enhance the building's entrance screen, which inspired this book's endpapers. Equally stunning is the patterned marble-floored central rotunda, with polished travertine walls, all surrounding the large pit where the Foucault Pendulum continues on its mysterious path that demonstrates the earth's rotation. This is one of the Observatory's most popular exhibits.

The pendulum is suspended from the center of a vaulted dome decorated by the celebrated muralist Hugo Ballin, whose mural displays various aspects of classical celestial mythology.

On the upper walls of the rotunda are eight rectangular murals also by Ballin, these depicting the advancement of science. Beyond the rotunda is the South Gallery, its ceiling dominated by a mural with a magnificent abstracted sun motif, painted by muralist, Anthony B. Heinsbergen, whose moated French chateau-style studio building still exists on Beverly Boulevard.

There can be no finer way to spend a day in Los Angeles than to visit Griffith Observatory. The elegance of the architecture is equaled by the fascinating displays of astronomical treasures within. Spread out all around is the commanding vista from the mountains to the sea, where at night the city lights attempt to compete with the stars above – a many-splendored thing.

Arnold Schwartzman
Hollywood, 2015

Sidewalk terrazzo medallion depicting Los Angeles City Hall. Architects: John Parkinson, in collaboration with John Austin. Clifton's Brookdale Cafeteria, 1936.

Col. Griffith Jenkins Griffith, Philanthropist

Griffith J. Griffith was born in South Wales in 1850, and grew up on a farm under very poor circumstances. An uncle visiting from America offered to take Griffith back with him. Griffith enthusiastically accepted and in 1866 sailed to New York on *The City of London*. First settling in Ashland, Pennsylvania, Griffith moved to San Francisco, California, in 1873, where he became the manager of the Herald Publishing Company.

Five years later, he met Fredrick McCrillish, the owner of the *Alta California* newspaper, who offered Griffith a job as the paper's mining correspondent.

Although the Gold Rush of 1849 was long past, San Francisco was still seen as a boomtown.

As a reporter, Griffith gained extensive knowledge of the mining industry, which eventually led to his accumulation of a large fortune.

In 1882, Griffith moved to Los Angeles, where he invested in real estate, including a large area of the Rancho Los Feliz Mexican land grant. In 1896, he donated 3,015 acres of this land to the City of Los Angeles to establish a public park, named in his honor, and told the City Council, "I consider it my obligation to make Los Angeles a happy, cleaner, and finer city. I wish to pay my debt of duty in this way to the community in which I have prospered."

Griffith underwent a violent personality change induced by his heavy drinking.

On September 3,1903, while on vacation, Griffith got into a dispute with his wife and shot her. He was charged with assault with a deadly weapon with intent to commit murder and subsequently spent two years in San Quentin State Prison. Griffith became the first defendant to benefit from a reduced sentence due to being impaired by his addiction. He stopped drinking, and devoted himself to prison reform and improving Griffith Park.

During a visit to Mount Wilson in 1908, Griffith was given the opportunity to peer through the observatory's 60-inch telescope.

It was said that in an attempt to foster goodwill, inspired by his celestial experience on Mount Wilson, in 1912 Griffith pledged an additional gift to Los Angeles of $100,000 to build an observatory and theater on the park's highest peak, Mount Hollywood, which he referred to as a "Christmas present."

The offer was accepted by the city council. Still perceived as not a very likeable person, members of the park commission objected and instituted a court action to block the donation.

Griffith later donated another thousand acres along the Los Angeles River.

Griffith died on July 6, 1919. In his will, he bequeathed the bulk of his $1.5 million estate to a trust that would build the Greek Theatre (1929) and Griffith Observatory (1935), and on completion, the buildings were to be donated to the City of Los Angeles.

John C. Austin, Architect

John Corneley Wilson Austin was born in Bodicote, Oxfordshire, England in 1870. In 1888, he left for New York, then moving on to San Francisco, where he lived for several years. Due to the economic crash of 1893, Austin returned to England. He came back to the U.S. in 1894 and settled in Los Angeles, where he remained for the rest of his life.

Austin became a prominent architect as well as a respected civic leader. He participated in the design of several Los Angeles landmark buildings, including Los Angeles City Hall, the Shrine Auditorium, NBC Radio City, and Griffith Observatory.

In partnership with Frederic Ashley, he began work on the Griffith Observatory in 1931. Their design was based in part on the drawings by Russell W. Porter.

The building, designed in the Art Deco style of the period, incorporates a number of "Deco Greco" Greek Revival architectural features.

As a result of the March 10, 1933 Long Beach earthquake, Austin and Ashley altered their original plans. The structure was reinforced with thicker concrete and additional steel. They dispensed with the building's planned exterior terra-cotta sheathing and ceramic tiles on the planetarium dome and replaced them with copper panels.

During the Great Depression, prices for building supplies were low, so the architects were able to purchase some of the most elegant materials for the building's interior, such as travertine marble, the finest wood, and bronze metalwork.

The official groundbreaking took place on June 20, 1933, and the Observatory's grand opening was on May 14, 1935. John Austin died in 1963.

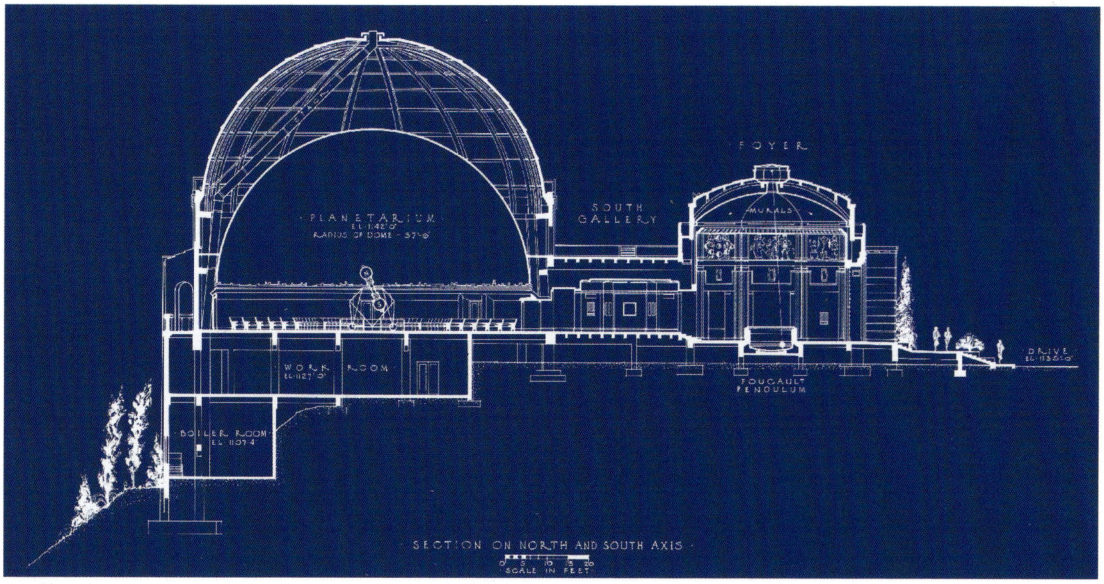

John C. Austin and Frederic M. Ashley's Observatory architectural plan, section of north-south axis.

Little acknowledgement has been afforded to Russell W. Porter (1871-1949).

Porter was a Renaissance man with many diverse interests, including art, architecture, polar exploration, navigation, music, and telescope-making. The least-known contribution was his fundamental role in the design of Griffith Observatory.

A consultant to Austin and Ashley's architectural practice, Porter's design drawings played a key role in the early conceptual plans for the building.

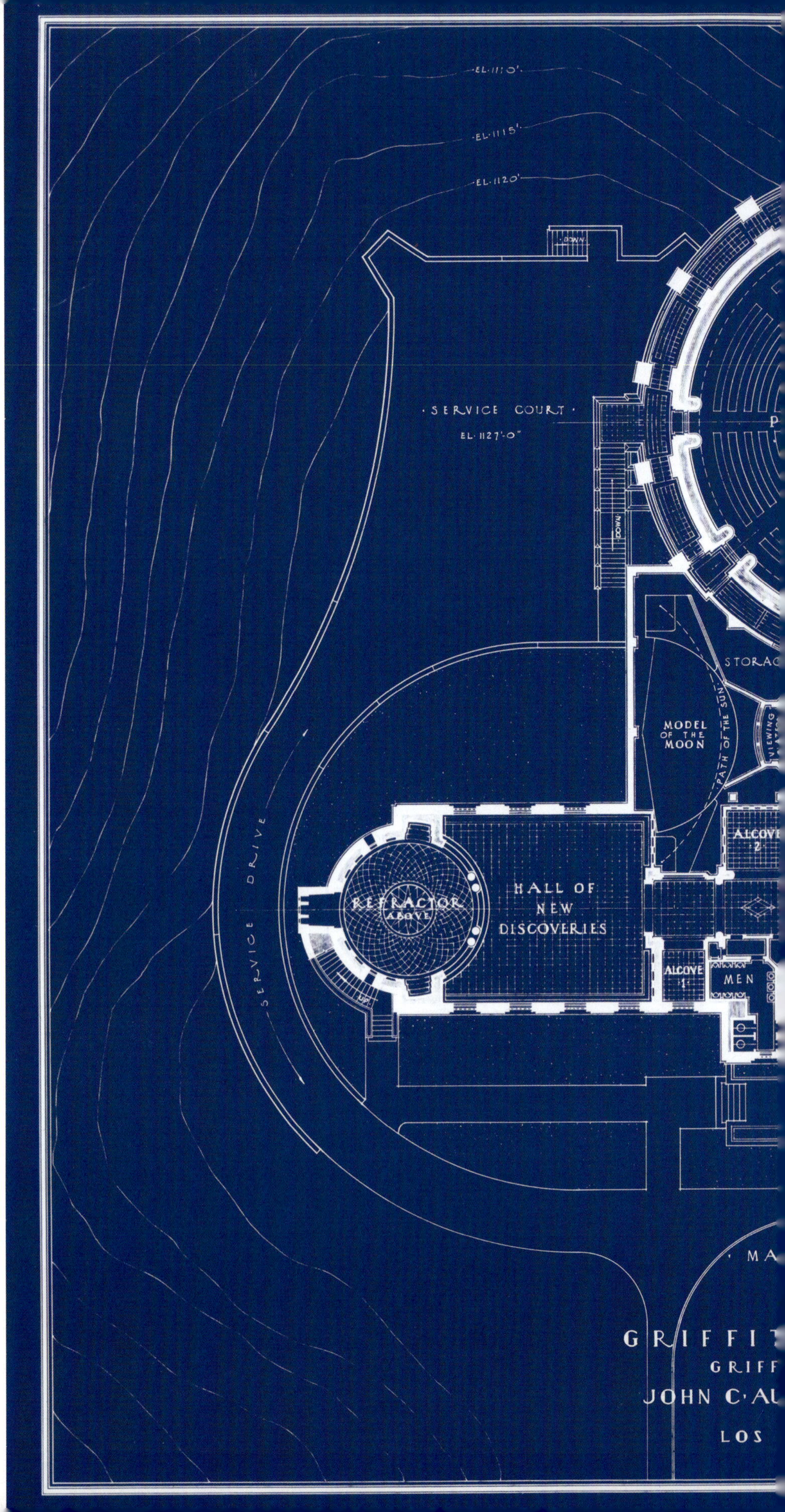

Griffith Observatory architectural main floor plan by John C. Austin and Frederic M. Ashley.

Pages 22–23: Conceptual renderings by John C. Austin and Frederic M. Ashley for Griffith Observatory.

Rendering by N.W. Alpaugh for architects Austin & Ashley's proposed design for Griffith Observatory, 1931.

W. W. ALPAUGH 31
OBSERVATORY AND
HALL OF SCIENCE
GRIFFITH PARK.
LOS ANGELES CALIFORNIA.
JOHN C AUSTIN FAIA & FREDERIC M ASHLEY AIA
ARCHITECTS

Rendering by
N.W. Alpaugh,
1931.
Alpaugh was in
partnership with
A.E. Sedgwick
in a Los Angeles
architectural practice.

OBSERVATORY AND
HALL OF SCIENCE
GRIFFITH PARK
LOS ANGELES CALIFORNIA
JOHN C AUSTIN · FAIA · & FREDERIC · M · ASHLEY AIA
· ARCHITECTS ·

Skeletal frame of the planetarium dome, February 1934.

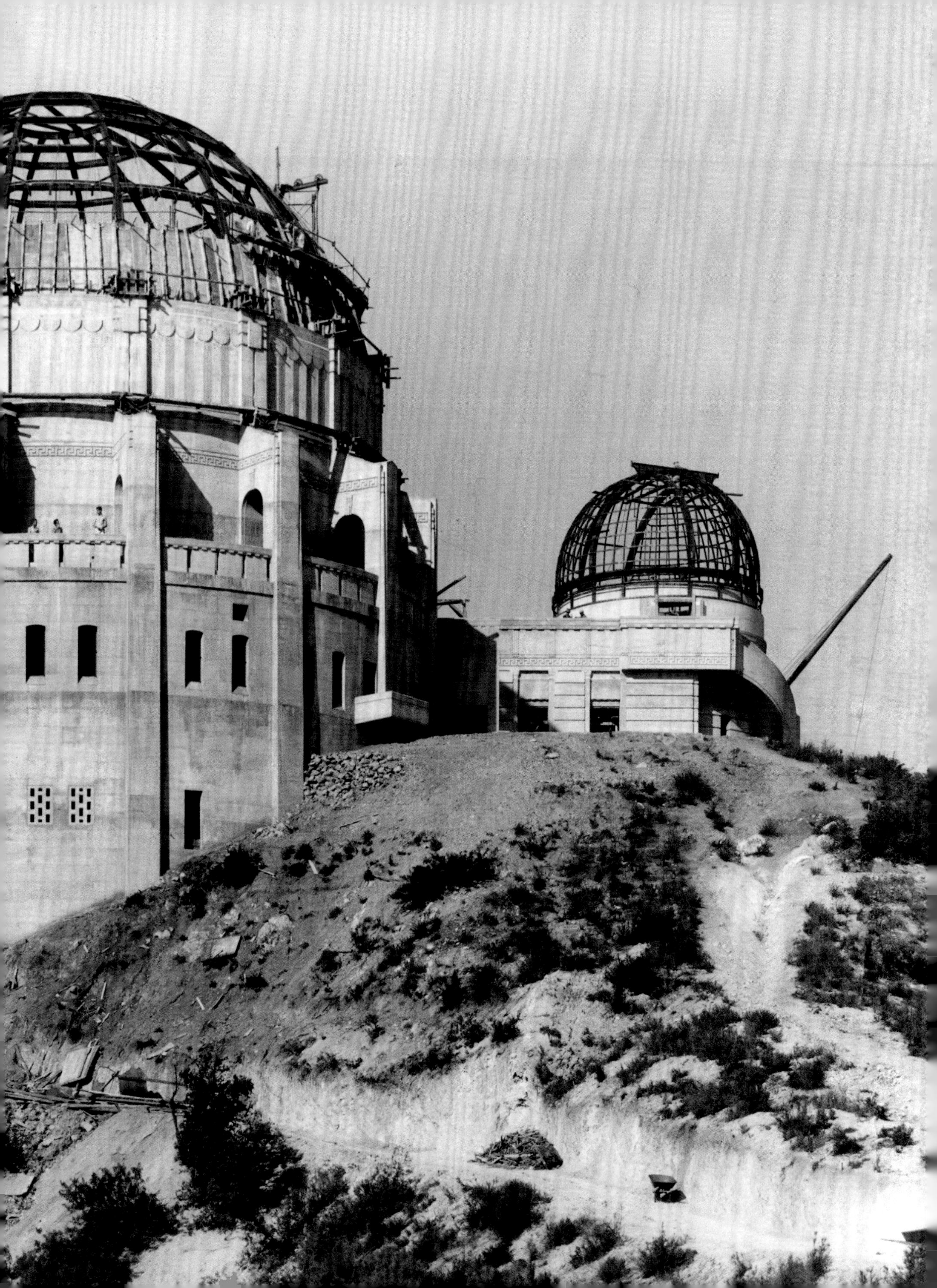

Facade of the Observatory during construction.

Quotation, right: A remark made by Colonel Griffith after peering through Mount Wilson Observatory's 60-inch telescope, 1908. Griffith Observatory and plaza close to completion.

Note Los Angeles City Hall, the tallest downtown structure, in the distance on the left.

"If all mankind could look throug

hat telescope, it would change the world."

— GRIFFITH J. GRIFFITH

Archival aerial view of the operational Observatory.

Astronomers Monument and Griffith Observatory facade.

Pages 34-35: View of the Observatory from the south, looking north.

COPERNICUS

Astronomers Monument

The 40-foot-high, 25-foot-wide, fluted obelisk monument honors the legendary pioneers in astronomical achievements:

Hipparchus of Nicaea, Nicolaus Copernicus, Galileo Galilei, Johannes Kepler, Sir Isaac Newton, and Sir William Herschel. The 9-foot-tall figures of the astronomers are arranged around the hexagonal spire.

The monument was funded through the New Deal initiative's Public Works of Art Project (PWAP) and the Women's Auxiliary of the Los Angeles Chamber of Commerce.

The PWAP sponsored a competition for the design, and on March 13, 1934, the Los Angeles Parks Commission selected the design by Archibald Garner.

Five other local sculptors were then chosen to create the figures in keeping with Garner's Art Deco design. Which sculptor carved each astronomer is unknown.

The sculptures of the astronomers are constructed in cast concrete and now topped with a cast bronze armillary sphere. All work on the monument, including the bronze casting, was done on site.

The figures were first modeled in clay and then cast in plaster for final finishing, from which molds were made. These were incorporated into the wooden forms for the obelisk, and the entire monument was cast in one pour of concrete.

The dedication ceremony took place on November 25, 1934.

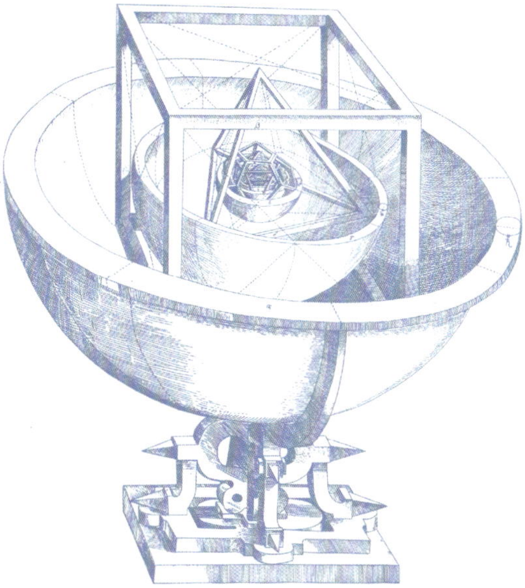

The "Harmony of the Planetary System" according to Kepler.

Opposite: Astronomers Monument, eastern aspect.

Hipparchus of Nicaea

Hipparchus, c.190–120 BCE, Greek, discovered the precession of the equinoxes and the eccentricity of the sun's path, determined the length of the solar year, estimated the distances of the sun and moon from the earth, drew up a catalogue of 1,080 stars, fixed the geographical position of places by latitude and longitude, and invented trigonometry.

Nicolaus Copernicus

Nicolaus Copernicus, 1473–1543, Polish, developed a heliocentric model of the solar system that explained planetary retrograde motion and overturned Greek astronomy.

Galileo Galilei

Galileo Galilei, 1564–1642, Italian, performed fundamental observations experiments, and mathematical analyses in astronomy and physics. He discovered mountains and craters on the moon, the phases of Venus, an the four largest satellites of Jupiter: Io, Europa, Callisto, and Ganymede.

ohannes Kepler	Sir Isaac Newton	William Herschel

hannes Kepler, 1571–1630, German, tablished the most exact astronomical bles of the seventeenth century. n astronomer, astrologist, and athematician, Kepler established e three laws of planetary motion.	Sir Isaac Newton, 1641–1727, British, developed theories of gravitation and mechanics, and invented differential calculus.	William Herschel, 1738–1822, British (German-born), discovered Uranus and its two brightest moons, Titania and Oberon; two of Saturn's moons, Mimas and Enceladus; the ice caps of Mars; several asteroids and binary stars; and cataloged 2,500 deep-sky objects.

The Monument Sculptors

Archibald Garner, 1904–1969

Garner was born in Onida, South Dakota. His family moved to the Pacific Northwest and then southward, ending up in Los Angeles. He graduated from Long Beach High School in 1922, and later studied at the Chouinard Art Institute.

Garner's work includes his wood bas-relief *Centinela Springs* at the Hillcrest Inglewood Post Office,1937, and his 1941 limestone female figure of *Law* for the Los Angeles U.S. Federal Courthouse lobby. He also worked as a sculptor for 20th Century-Fox Studios.

Roger Noble Burnham, 1876–1962

Roger N. Burnham graduated from Harvard College in 1899, and later produced decorative sculptures for the 1904 St.Louis World's Fair. He moved to Boston, where he continued to work as a sculptor, specializing in portraiture and designing medals. In 1922, he moved to Los Angeles, where he taught modeling at Otis Art Institute. He was locally renowned for his 1930 sculpture of *Trojan* (aka Tommy Trojan) at the University of Southern California. That same year, he sculpted *Aspiration* in memory of silent film idol Rudolph Valentino. It was film star Dolores del Rio, wife of famed movie art director Cedric Gibbons.

Aspiration, statue in memory of Rudolph Valentino, 1930. De Longpre Park, Los Angeles. Artist: Roger Burnham.

Opposite: Detail, Astronomers Monument, dedicated 1934.

Left: *The Vanquished Race*, figure of a Native American, at Thomas Starr King Middle School, 1938. 4201 Fountain Avenue, Hollywood. Sculptor: Djey el Djey.

Right: Bronze statue of Beethoven, 1932. Pershing Square, Los Angeles. Artist: Arnold Foerster.

Gordon Newell, 1905–1998

Born in California, Gordon Newell studied at Occidental College in Los Angeles and University of California, Berkeley and then was apprenticed to Ralph Stackpole for work on the San Francisco Stock Exchange. Newell moved to Los Angeles, where he taught at the Chouinard Art Institute and Occidental College, and later taught at the Sculpture Center in Monterey.

Newell created (with Sherry Peticolas) the plaster facades for the Hillcrest Inglewood, California, Post Office in 1937. Newell was married to actress Gloria Stuart.

Djey el Djey, aka Djey Owens, 1905–1980

Owens was a resident of Los Angeles and active participant on the federal art project (PWAP). His sculpture, *The Vanquished Race,* is installed at Thomas Starr King Middle School, Hollywood.

Arnold Foerster, 1878–1943

The Astronomers Monument was first suggested by Foerster. He sculpted one of the figures, and as he was experienced in the making of large scale works, was designated to engineer the casting of the monument.

Foerster also created the bust of Col. Griffith for the Observatory in 1935.

In 1932, he created the 12-foot-high bronze *Beethoven* statue in Pershing Square Park, Los Angeles.

George Maitland Stanley, 1903–1970

George Stanley studied sculpture at the Otis Art Institute in Los Angeles (1923–1926) and taught there from 1926–1942. In 1927, the newly formed Academy of Motion Picture Arts and Sciences chose Stanley to sculpt a statuette based on a sketch believed to have been made by MGM art director Cedric Gibbons. This became the Academy Award statuette, popularly known as "Oscar."

In 1940, Stanley completed *Muse of Music, Dance, Drama,* the sculpture for the entrance to the Hollywood Bowl that is carved from granite, 22 feet tall and 200 feet wide. In 1928, he created the bas-relief over the Wilshire Boulevard entrance of the renowned Bullocks Wilshire department store building.

Bas-relief, Wilshire Boulevard entrance of the former Bullocks Wilshire department store, 1928-29. 3050 Wilshire Boulevard, Los Angeles. Sculptor: George Stanley.

George Stanley was responsible for sculpting Sir Isaac Newton on the Astronomers Monument.

The new armillary sphere on top of the Astronomers Monument replaced the original in 1991.

Right and opposite: Observatory sundial, 1957. The plaque at the base reads "In sun light the shadow of the wire indicates correct watch time on the ring."

Pages 54-55: Observatory facade with bronze entry screen of graphic symbols.

SUNDIAL
IN SUN LIGHT THE
SHADOW OF THE
WIRE INDICATES
CORRECT WATCH
TIME ON THE
RING.

GRIFFITH

SERVATORY

EXIT
EXIT

ENTRANCE
ENTRANCE

Two concrete bas-reliefs flank each side of the Observatory's north doors, representing Alpha and Omega. They are believed to have been created by Joseph Conrad, (under contract to architects Austin and Ashley).

Opposite: Detail of bronze graphic motif on north door grille.

Pages 60-61: Detail, bronze entry screen.

Archival photograph of visitors observing the Foucalt Pendulum in the Observatory's Central Rotunda.

Opposite:
Central Rotunda, looking toward the South Gallery and doors to the Samuel Oschin Planetarium.

EVENT HORIZON THEATER
SAMUEL OSCHIN
PLANETARIUM

The 240-pound brass ball which moves back and forth on a 40-foot steel cable suspended from Hugo Ballin's painted vaulted ceiling is a powerful illustration of Earth's rotation and its relationship with the heavens. At Los Angeles's latitude, it takes forty-two hours for the pendulum to swing completely through 360 degrees as Earth rotates. For decades, that movement has been visually represented by small wooden dowels that are knocked over one at a time by a pointer on the bottom of the shiny brass ball.

31
30
29
28
Pendulum
turning of the earth, and once started in motion, it continues to swing in the same direction regardless of what the earth does. The earth rotates beneath it, making it look as if the pendulum is changing its direction of swing.
At the North and South Poles, it takes 24 hours for a complete circle of pegs to be moved into the pendulum's path. The closer a pendulum is to the equator, the more slowly a peg is brought into its path. At the equator itself, there is no apparent shift, and no pegs would be knocked down. In Los Angeles, the pendulum's apparent direction of swing rotates through a full circle in a little more than 42 hours, so the dial in the pit is divided into 42 hourly divisions.

The Muralists:
Hugo Ballin and Anthony Heinsbergen

Hugo Ballin was born in New York City in 1879 and studied there at the Art Students League, as well as in Italy and France.

One of the nation's premier and prolific muralists, Ballin was also an author, film director and producer, and is credited with producing and directing over one hundred silent movies. He was a veritable polymath.

Of the numerous and admired murals he created throughout Los Angeles, his murals for Griffith Observatory are proclaimed to be among his finest.

The Observatory's Keck Central Rotunda's vaulted ceiling commemorates astronomical mythology with images of Atlas, the four winds, the planets as gods, and the twelve constellations of the zodiac.

His eight 6-foot-by-11-foot wall murals depict the "Advancement of Science" with panels on time, navigation, astronomy, geology and biology, metallurgy and electricity, civil engineering, mathematics and physics, and aeronautics.

In 1917, Samuel Goldwyn hired Ballin as an art director for MGM studios in New Jersey. He relocated to Los Angeles in 1921, formed his own production company, and directed costume dramas including *Jane Eyre* (1921), as well as designing sets for Cecil B. DeMille. He later provided sketches for *The Wizard of Oz*.

In 1929, another team of Hollywood moguls, the Warner brothers, commissioned Ballin to paint the Wilshire Boulevard Temple's 320-foot-long pageant of Jewish history-themed Warner Murals.

In addition to his noted murals at the Wilshire Boulevard Temple, his other works include: the Los Angeles Times Building – the ten-foot-high murals surrounding the building's rotunda depict the sources and carriers of news; Los Angeles City Hall Council Chambers; the Title Insurance and Trust Company Building (1928); and the Southern California Edison Company Building (1930).

Ballin was chosen to design the Los Angeles Memorial Coliseum's peristyle's giant medallion for the 1932 Los Angeles Olympic Games.

Opposite and Pages 68-75: Details, Hugo Ballin's vaulted ceiling mural depicting mythology. Griffith Observatory central rotunda, 1934–35.

Below: Hugo Ballin's giant commemorative medallion (extant), *Pallas Athena*, deity of the Olympic Games, with motto: *Citius Altius Fortius* (Swifter Higher Stronger) for the 1932 Los Angeles Olympic Games. Los Angeles Memorial Coliseum, 3911 S. Figueroa Street, Los Angeles.

·HVGO·BALLIN·
·1934·

Pages 76–83: Hugo Ballin's eight 6-foot-by-11-foot wall murals surround the W.M. Keck Foundation Central Rotunda.

Navigation

On the left, a symbolic figure of the Wind grasps a sundial and compass. Calm, shown to the right, is holding a modern sextant. The moon orbits diagrammatically around the two figures. At the top is an astrolabe, an early astronomical instrument.

Metallurgy and Electricity

On the left, St. Florian, patron of metallurgists, sits, pouring water over fire beneath a converter for steel-making. A kite and key recall Benjamin Franklin's discovery of lightning's electrical charge.
Otto von Guericke holds an early device which produced electricity through friction.

© 1934

Astronomy

On the left, Arzachel consults his early astronomical tables. John of Holywood works on an influential astronomical text. Copernicus, who proved that the sun, not Earth, was the center of the solar system, stands near Galileo, whose telescope is dwarfed by a modern instrument.

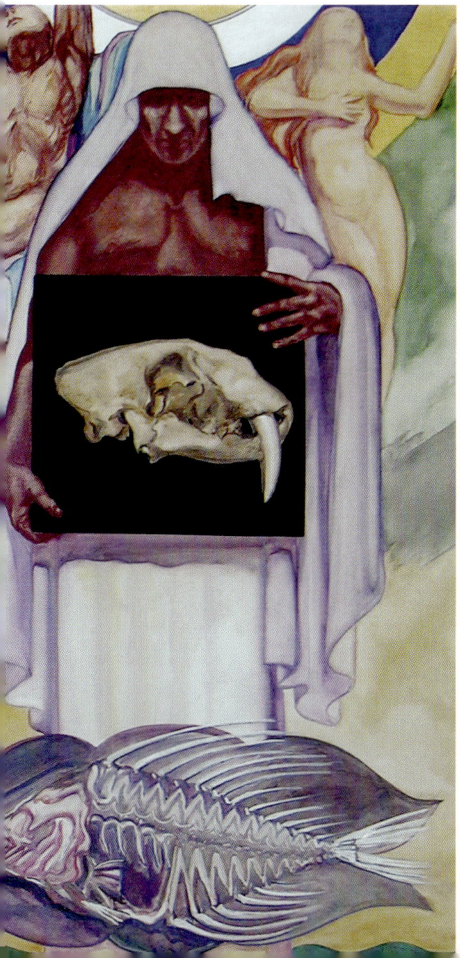

Geology and Biology

An old man symbolic of geology holds a card showing mineral crystals. Through a microscope, a biologist studies life, with a chicken embryo, plant, and crab below. A man symbolizing paleontology shows a saber-toothed cat skull with a fish skeleton below.

Time

To the left an Aztec points to the ancient Calendar Stone. Emperor Yao of China witnesses the execution of magicians who failed to prevent a solar eclipse. On the right is Ulugh Beg, grandson of Timur and keen celestial observer.

Mathematics and Physics

On the left is a Greek geometer, whose work derived from the Egyptians. Their influence is represented by the Egyptian god Thoth, behind. An Arab looks at the important mathematical contributions of his culture. The seated man looking into the murky cloud represents physics. Sir Isaac Newton, originator of calculus, talks with students on the right.

Aeronautics

The artist portrays Archytas of Tarentum with his wooden pigeon, Roger Bacon with his floating metal ball, Francesco de Lana with a flying boat model, Leonardo da Vinci, whose models perhaps needed only an engine, and Besnier, gliding successfully in 1678. Behind Besnier is a Boeing 247, the first modern passenger airliner, introduced in 1933.

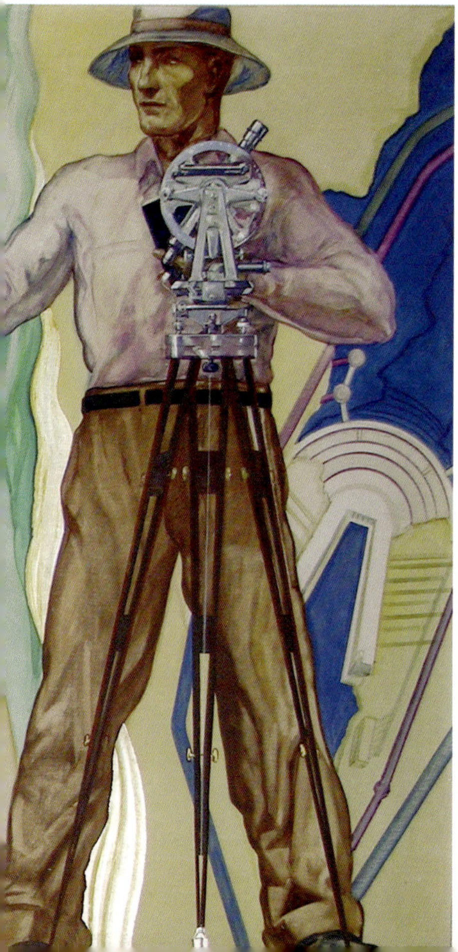

Engineering

On the left, an ancient Egyptian stands before a pyramid, an impressive monument to ancient engineering. To the left of the Egyptian is the star Sirius, associated with the flooding of the Nile river. Two grappling figures represent forces within the earth associated with inevitable earthquakes that challenge the work of engineers. Modern engineering is represented by a surveyor and Hoover Dam.

Anthony B. Heinsbergen

Anthony B. Heinsbergen, born in Holland in 1894, immigrated to Los Angeles in 1906 and continued his art education at Chouinard Art Institute. He became firmly established among America's greatest interior designers, notably for his ornamental decoration of the Palace Theatres.

An accomplished muralist, he founded the Heinsbergen Decorating Company in 1922 and went on to hire more than a hundred artists. He is also credited with decorating over 750 theaters nationwide.

Among Heinsbergen's other notable commissions in the Los Angeles area are the Union Trust & Savings Bank, Hollywood Roosevelt Hotel, the Elks Club, the Gables Beach Club in Santa Monica, Tower Theatre, Wiltern Theatre, Los Angeles City Hall, and the Beverly Wilshire Hotel. Anthony Heinsbergen died in 1981.

Opposite and pages 86-87: The Observatory's East Gallery ceiling decoration depicts the sun and its burgeoning rays. Note: the circular key pattern was not completely rendered.

Below: Decoration above the entrance to the former Heinsbergen Studio, 7415 Beverly Boulevard, Hollywood.

Lamp fitting,
South Gallery.

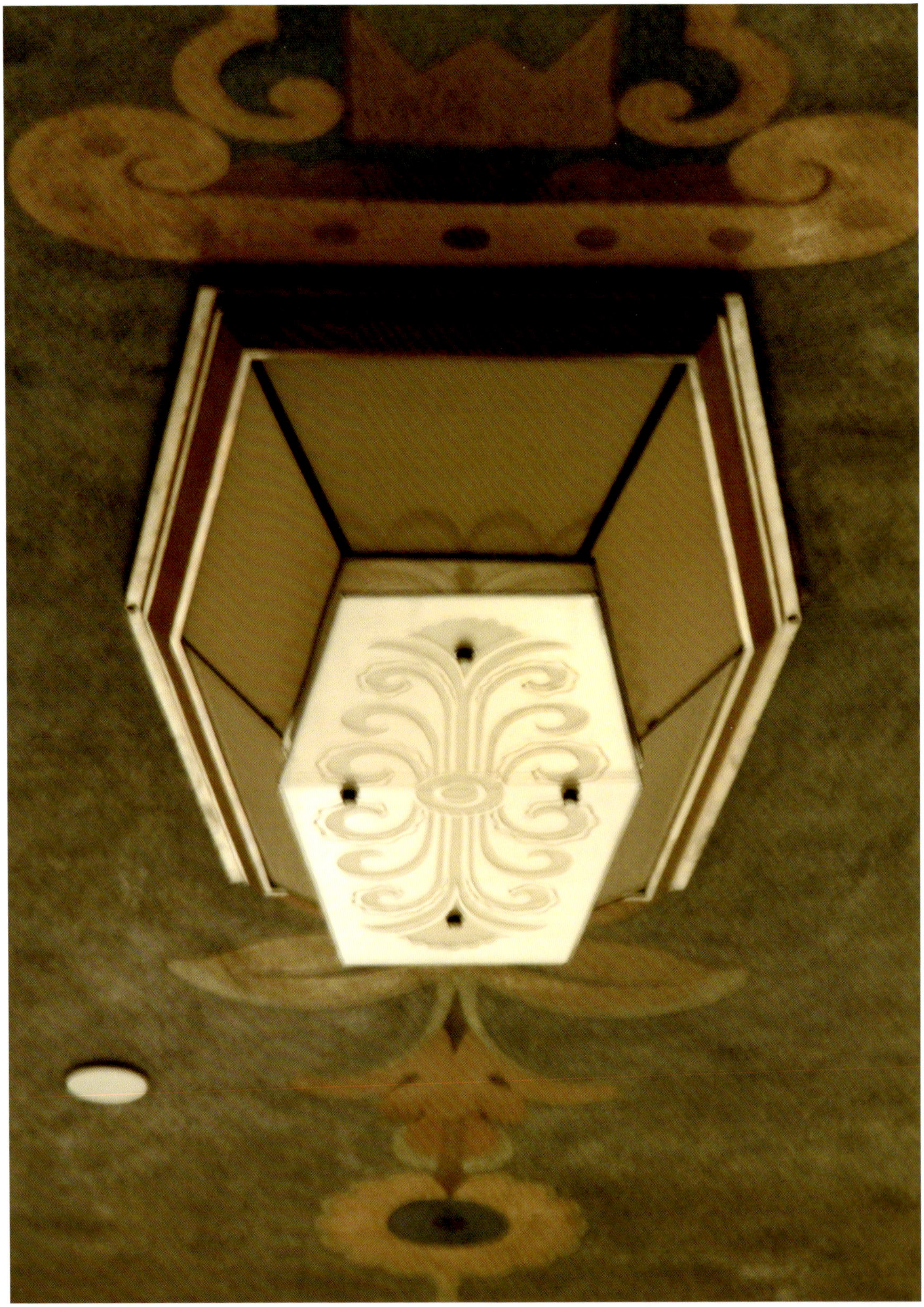

Detail, South Gallery.
Designer:
Anthony Heinsbergen.

Clock,
W.M. Keck Foundation
Central Rotunda.

Opposite: Inlaid brass ancient Greek directional signs.
Top: West.
Bottom: East.
W.M. Keck Foundation Central Rotunda marble floor.

ΖΟΦΟΣ

ΗΩΣ

Metal
ventilation grille.

Opposite:
W.M. Keck
Foundation
Central Rotunda
with Foucault
Pendulum well.

INFORMATION
TICKETS
EXIT
NOTOS

Right and pages 96-97: Observatory north entrance and copper-clad west dome housing the coelostat telescope.

ATORY

GRIFFITH
INFORMATION

BSERVATORY
INFORMATION

TO
TELESCOPE

Detail, copper acanthus leaves motif, exterior cupola surrounding W.M. Keck Foundation Central Rotunda.

Opposite:
Illuminated sign to the 12-inch Zeiss refractor telescope.

Pages 100-107: The West Terrace and the solar telescope's dome.

HOLLYWOOD

The Samuel Oschin Planetarium dome, view towards the Pacific Ocean.

Pages 110-111:
Detail, decorative fluting, Samuel Oschin Planetarium dome.

Gathering storm over Griffith Observatory's copper-domed Samuel Oschin Planetarium and the Promenade Walkway with downtown Los Angeles beyond.

← TO TELESCOPE

Directional sign to the 12-inch Zeiss refractor telescope, Roof Terrace.

Dinsmore Alter, Observatory Director (1935-58), demonstrates the original Zeiss Mark II planetarium projector to the "Welcomettes."

Opposite:
The Zeiss 12-inch refractor telescope.

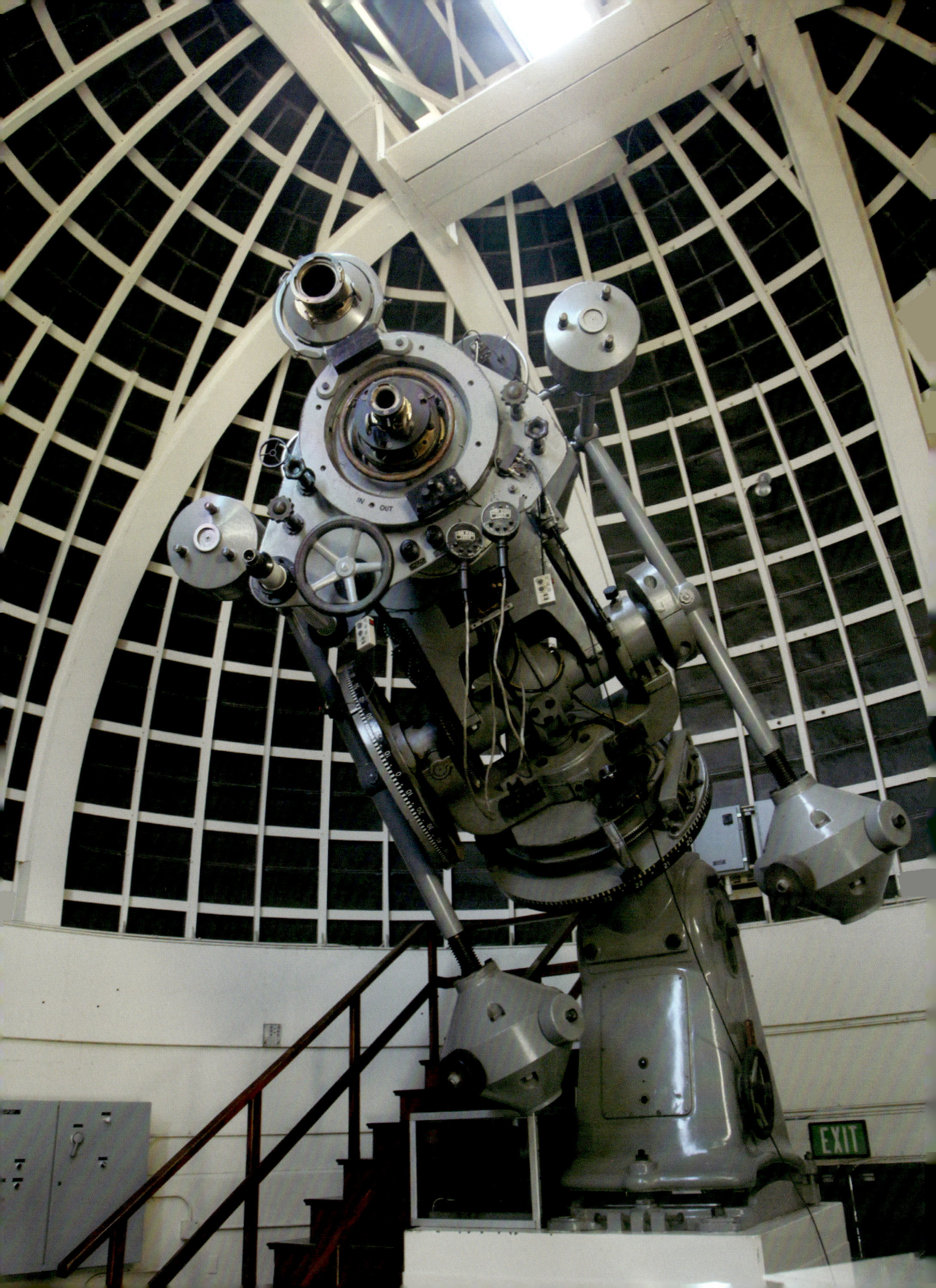
IN
OUT
EXIT

Sunset over the Hollywood hills and the Hollywood sign framed by an arch of Promenade Walkway.

Opposite: Astronomers Monument at sunset.

Pages 120-121: The barrel of the 12-inch Zeiss refractor telescope visible in the open roof.

Sunset on the Astronomers Monument.

Griffith Observatory viewed from Mount Hollywood Trail.

Sunset on the Astronomers Monument.

GRIFFITH

RVATORY

A waxing crescent moon to the southwest of the Observatory.

Sunset,
Astronomers
Monument.

Opposite: Detail of Observatory's bronze north exit door.

Griffith Observatory:
A Celebration of its Architectural Splendor

Photographs and text

ISBN 978-0-940512-54-2

Library of Congress Cataloging-in-Publication Data
Schwartzman, Arnold.
Griffith Observatory : a celebration of its architectural splendor / photographed by Arnold Schwartzman.
pages cm
ISBN 978-0-940512-54-2 (pbk. : alk. paper)
1. Griffith Observatory--History. 2. Griffith Observatory--Pictorial works. 3. Astronomical observatories--California--Los Angeles--Design and construction--Pictorial works. I. Title.
QB82.L87 S33 2015
522'.1979494--dc23
2014042035

Design by Arnold Schwartzman
Art production by Isolde Schwartzman

HENNESSEY + INGALLS
ART & ARCHITECTURE BOOKSTORE
Published by Hennessey + Ingalls
214 Wilshire Boulevard
Santa Monica, California 90401
www.hennesseyingalls.com

Printed in Canada

Acknowledgments

With special thanks to my wife and partner, Isolde, for the book's art production.

The author would like to thank the following for their invaluable and generous assistance in preparation of this book:

Councilmember Tom LaBonge
Los Angeles City Council, Fourth District

Kamilla Blanche, Director
Sister Cities and Arts and Culture

Dr. E.C. Krupp, Director
Griffith Observatory

Mark Pine, Deputy Director
Griffith Observatory

Anthony Cook, Astronomical Observer
Griffith Observatory

Camille Lombardo, Executive Director
Friends of the Observatory

Mark Hennessey

Paddy Calistro
Scott McAuley

Archival photographs courtesy of:
Griffith Observatory Archive
Los Angeles Public Library Photo Collection
Marc Wanamaker, Bison Archives

Thomas Starr King Middle School

EXIT